ECOLOGICAL PACKAGING

Abstract

Global warming is one of the most concerning issue across the world. Mostly human being are responsible for this global warming but the entire planet is the victim by its adverse effects. There are many ways that are destroying our environment every day. Increasing plastic products is one of those, such as plastic shopping bags. Therefore, sustainable development is the biggest topic in the business corporate world. The Human being can play a vital role in the sustainable development in various ways regardless their own views.

This study researches on creating awareness to the modern society regarding our ecology and environmental aspects. Besides, the awareness will encourage people to use eco-friendly shopping bags (BioBag), instead of plastic shopping bags. While people will get used to the using of BioBag, certainly the manufacturing company will be benefited and continue business in a long period of time.

The case company named ''BioBag® World''. Basically, it is a Norwegian company and production units and market areas are in Europe and North America as well. In addition, a survey was carried out on different age groups and professions of the people. This thesis work determines various opportunities and aspects of sustainable development and environment. This study examines global warming, sustainable environment, and the strategies of various organizations, the consumer behavior and marketplace and future opportunities of BioBag as shopping bag in Finland as well as all over the world.

The outcome of this thesis shows that the surveyed group of people liked BioBag as a shopping bag instead of plastic shopping bag. Moreover, the people are completely agreed that the BioBag is 100% environmental friendly than the plastic bag. The people also think that the can also use BioBag for different purposes and it is also enough durable bag. It would be advisable to the case company to pay more attention the marketing and providing more information to the consumers regarding BioBag

LIST OF TABLES AND FIGURES

TABLES

Table 1. Difference between plastic shopping bag and BioBag shopping bag

FIGURES

Figure 1. Yearly carbon emission globally

Figure 2. Global trends in major green house gases 1/2003

Figure 3. Dimension of sustainable development (IISD 2010)

Figure 4. Parts of Three dimension of sustainable development

Figure 5. Adverse effects of using Plastic throughout the world

Figure 6. Life cycle process of BioBag (shopping Bag)

Figure 7. Mater-Bi®- Raw materials of BioBag

Figure 8. Samples of BioBag (shopping bags) made by BioBag Company

Figure 9. Gender based BioBag users (n=20)

Figure 10. Age group of the user (n=20)

Figure 11. Profession of the user (n=20)

Figure 12. Plastic bag user per month (n=20)

Figure 13. Idea and care about global warming (n=20)

Figure 14. Familiarity of Eco friendly bag (n=20)

Figure 15. Satisfactions about using BioBag (n=20)

Figure 16. Use of Plastic bags (n=20)

Figure 17. Business Idea of manufacturing BioBag (n=20)

Figure 28. Eagerness of using BioBag in everyday life (n=20)

Figure 19. Multiple use of BioBag (n=20)

Table of Contents

1 **INTRODUCTION** 8
   1.1 PURPOSE OF THE STUDY 8
   1.2 RESEARCH QUESTION 8
   1.3 OUTLINE OF THE STUDY 9
      1.3.1 Introduction to the research area 9
      1.3.2 Literature of the study 9
      1.3.3 Market area and discussions 9
      1.3.4 Conclusion and suggestions 9

2 **GLOBAL WARMING** 10
   2.1 REASONS OF GLOBAL WARMING 10
   2.2 RESULTS 12

3 **SUSTAINABLE DEVELOPMENT** 14
   3.1 DEFINITION 14
      3.1.1 Social 15
      3.1.2 Economical 16
      3.1.3 Environmental 17
   3.2 ADVERSE EFFECTS OF PLASTIC BAGS 18
   3.3 SUSTAINABILITY AND ECOLOGICAL ASPECTS OF PACKAGING 22
      3.3.1 Brief explanation of points 1-7 22

4 **CASE COMPANY PROFILE AND BIOBAG (SHOPPING BAG)** 25
   4.1 INTRODUCTION 25
   4.2 BIOBAG® (SHOPPING BAG) 26
   4.3 ENVIRONMENT (WE CARE) COMPOSTABLE AND BIODEGRADABLE 26
   4.4 LIFE CYCLE OF BIOBAG 27
   4.5 PRODUCTION AND MATERIALS COLLECTION 30
   4.6 FINAL PRODUCTS 31

5 **CONSUMERS** 33
   5.1 ECOLOGICAL CONSUMERS 33
   5.2 DIFFERENCE BETWEEN BIO BAG AND PLASTIC BAG 34
   5.3 CREATING AWARENESS AND ENCOURAGE PEOPLE TO BUY ENVIRONMENTAL FRIENDLY SHOPPING BAGS (BIOBAG) 35

    5.4   Environmental friendly products, price, promotion, place and purchasing 36

    5.5   Business and Eco-friendly marketing 38

    5.6   Advertising 39

6   RESEARCH METHODOLOGY 40

    6.1   Research Method 40

    6.2   Validity and reliability 40

    6.3   Results 41

        6.3.1   Preference based on gender 42

        6.3.2   Preference based on age group 43

        6.3.3   Profession 44

        6.3.4   Plastic bags users/month 45

        6.3.5   Idea and care about global warming 46

        6.3.6   Familiarity of Eco friendly bag 47

        6.3.7   Satisfactions about using BioBag 48

        6.3.8   Plastic bag after use 49

        6.3.9   Business Idea 50

        6.3.10   Eagerness of using BioBag in everyday life 51

        6.3.11   Multipurpose use of Biobag 52

        6.3.12   Summary of the results 53

7   CONCLUSION 53

    7.1   Summary of the theoretical framework 54

    7.2   Suggestion 55

    7.3   Suggestion for future research 56

REFERENCES 57

APPENDICES 60

1 INTRODUCTION

1.1 Purpose of the study

Due to massive industrialization, the environment pollution increases in an alarming rate. The human being is much more concern about the environmental issues. They are more interested saving environment by using ecological friendly products. Every day we are using uncountable products to fill up our daily life demands. Unfortunately, most of these are manufacturing by plastic or other eco-unfriendly materials. Therefore, the ecology is getting imbalance and we are facing different types of natural calamities. It is considered that only environmental awareness and availability of eco-friendly products would increase the using of environment friendly products more than before. The purpose of this study is to find out future business opportunities of ecological packaging ''The BioBag Shopping Bag''.

1.2 Research question

Once upon a time, people had less concern about environment and ecology, but present days the concern changes dramatically and turn towards saving the environment in a greater speed. Now it's time to spread this positive concern throughout the world and the benefits of using eco-friendly products. Indeed, only business can make it possible faster. Moreover, it can be also said that what kind of relation exists between saving ecology and formulating business simultaneously. On the other hand, business wants profit because without profit business can't exist. It is considered that coming future for the eco-friendly products will be very impressive for the customers and profitable for the company. In other words, the raw materials would be cheap and easy to get for a company, which will cut extra cost and bring much profit. So it would be very easy for a company to live a long time to make their customers happy and make the environment clean and fresh as well. So, this study will consider, "How bio-bag (shopping Bags) will bring long run benefit for the company from the environmental aspect."

1.3 Outline of the study

This research was carried out in the company that manufactures the Bio shopping bag in Helsinki during 2015 and 2016. The study outcome report has been divided into following parts-

1.3.1 Introduction to the research area

In this part, there is a brief discussion of the research has been done. This part is contained fundamental information on what is studied about as well as researched process.

1.3.2 Literature of the study

The summary of this part includes sustainable development, Sustainable or Ecological packaging (shopping Bag), Ecological aspects of packaging, Bio-material shopping bag and its advantages, global warming, Adverse effects of using plastic, relation between Plastic shopping bag and global warming, Creating awareness to the people which will increase the use of BioBag (shopping bag) for their everyday life.

1.3.3 Market area and discussions

Precisely, the Finnish and global market will be described shortly as a target market. Moreover, there will be included consumers in the market place, business and eco-friendly marketing, create awareness so that people inspire to buy eco-friendly shopping bag (BioBag), environment-friendly products, price, place, promotion, and purchasing. The enormous opportunities in Finnish market and the encouragement in the Environmental care is the core point of this part. So in the long run how a company will be benefited from the environmental aspect and economical aspect.

1.3.4 Conclusion and suggestions

This portion of the report settles the study with a short summary of the research analysis and delivers advanced study suggestions to make it more developed and high-level.

2 GLOBAL WARMING

2.1 Reasons of global warming

The basic reason of global warming is growing temperature, but there are causing the vast range of additional secondary effects such as thermal expansion, because of thermal expansion of the ocean sea levels are increasing. In addition to that melting of land ice. Green house effects are largely responsible for increasing earth temperature. Green house gas concentrations like water vapor, carbon dioxide (CO_2), methane and ozone. Moreover, burning fossil fuels produces the largest amount of green house gas in the earth atmosphere. Besides, the increase of human-made emission of green house gasses is responsible for global warming. Sunlight is the biggest source of light, heat, and energy. Everyday sunlight reaches our earth surface and a small portion is absorbed and warms the earth and rest of the portion radiated return to the air at a longer wavelength than the sunlight. On the other hand, green house gasses absorbed a portion of the longer wavelength from our atmosphere before they lost to space. This long wavelength heat warms our atmosphere very badly. Because of green house gas, the heat cannot return to space. It returns to the atmosphere as reflecting energy. These green house gas work like mirror. And the reflecting heat is much more dangerous for our ecology and this reflecting heat energy is called greenhouse effect. Everything is closely related each other. Moreover, it is almost impossible to identify that a specific gas causes a specific amount of greenhouse effects. *(Cause and effects for global warming), (n.d)*

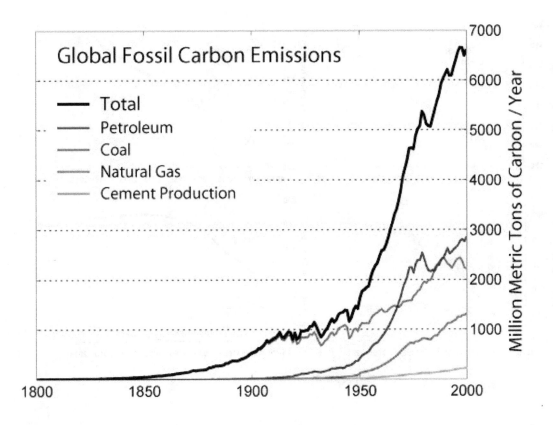

Figure 1. Fossil carbon emissions globally Graph: global warming art (source: *Fossil carbon emission, (n.d)*

According to the graph it is seen that from 1800 to 2000 the amount of CO2 increased and every year it is increasing in a high level.

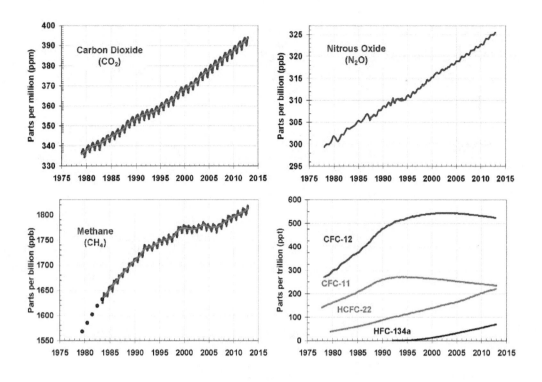

Figure 2. Trends in major green house gases, *Source: Trend of global gas emission (n.d),*

According to the graph, Carbon dioxide (CO_2) and nitrous oxide (NO_x) concentrations and all other gases in the atmosphere are increasing constantly.

2.2 Results

Due to global warming, our eco-system becomes destabilize. Many things go out of control such as sea level are rising, ice melting, numbers of natural disasters are increasing and all these things causing huge damage taking thousand of life every year.

Precisely, It can be said that all the information regarding global warming appears the most ignorant and inflexible. Here is some current knowledge regarding global warming:

- ❖ Surely, global warming exists and mainly caused by the human being. Moreover, it is still a continuous process.
- ❖ Because of global warming, we forecast that a rise of regular temperature which is responsible for among other things such as- melting of polar ice, glaciers, rising of sea level as well as extreme weather events and natural disasters such as- extreme rain, floods, drought, tornadoes and so on.

- Just a radical decrease of the waste gas emission can save the future generation and stop the trend as well.
- Indeed, the main causes of global warming belong to carbon dioxide emission. The extreme CO2 produced while burning fossil fuels like oil, natural gas, diesel, organic diesel, petrol, organic patrol, ethanol all is responsible for producing CO2.
- There was a time when many developed countries had a straight connection between energy consumption and welfare of a country. But now a day some countries have taken serious action to reduce environmental pollution. For example, Germany is reducing environmental pollution and keeping their economic growth high.
- Moreover, France has taken several strong steps to stop using plastic shopping bags as well as other products.
- Nordic countries are already so much sensitive regarding the green environment. Norway could be the best example for the world. In 2016 they have decided, Norway would ban all fuel or gas burning cars within 2025.

According to Kyoto Conference, many countries have agreed to decrease their CO2 emission until the year 2012, on average to 5% below their respective emissions of the year 1990. Moreover, it is known that USA is the largest CO_2 producer around the world as a result, most developing countries denied to sign this agreement as well.

3 SUSTAINABLE DEVELOPMENT

3.1 Definition

Today, sustainable development is one of the biggest discussed issues in the world. There are many definitions regarding sustainable development. In 1987 first sustainable development definition appeared-

"Development that meets the needs of the present without compromising the ability of future generations to meet their own needs."

— From the World Commission on Environment and Development's (the Brundtland Commission) report Our Common future. (Oxford: 1987).

Although, the definition sounds so simple to us but it is not that easy to understand. To understand its inner meaning we must go through some substances. These substances can be found our daily life activities and ask our self that- is there any need that conflict with each other? Then we will figure out many conflicts. For Instance, we want to get fresh air to breath but we need cars for better transportations, companies need cheaper workers whereas workers need for livable wages when people's need firewood and our ecology needs more trees to balance ecosystem. One country has mills and factories for development, on the other hand, it causes acid rain to other country's lakes, rivers etc. Sustainable development is one of the complex concepts for us. But there is three ways to take decisions and make action about sustainable future:
(Definition and pillars of sustainable development b)

1. Social
2. Economical
3. Environmental

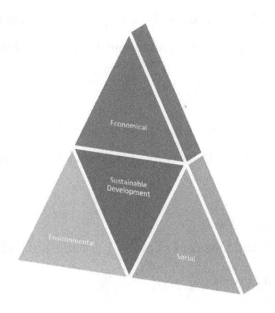

Figure 3. Three dimension of sustainable development. (IISD 2010.)

Social	Economical	Environmental
• Equity • Participation • Empowerment • Social mobility • Cultural preservation	• Services • House hold needs • Industrial growth • Agricultural growth • Efficient use of labor	• Clean air and water. • Biodiversity • Natural resources • Eco system integrity

Figure 4. Parts of Three dimension of sustainable development, (source: IISD2010.)

3.1.1 Social

It's a challenge to identify social pillars. A mass of concepts is connected to the term of "social". The oxford concise English dictionary provides about seven definitions mention for social.

Littig & Greissler (2005) note that the term has both analytical and normative meanings. There are also some complexities about the identification of 'Authentic' social issues, as significant overlaps exist across the three pillars of sustainable development.

According to the OECD 2009, this overlap is particularly admitted with respect to the social & economic (Thin, 2002) with many issues specially employment and unemployment, deemed relevant to both dimensions. (OECD, 2009)

Instead of these circumstances, the literature describes some policies that have been identified as 'social' within the sustainable development framework. These have been pronounced in various ways as:

- Social categories UNDESA, 2001
- Social dimensions OECD, 2009 and Dempsey et al 2011, Casula Vifell & Soneryd 2012;
- Social indicators UNCSD, 1996; UNDESA, 2001
- Social realm (Chan & Lee, 2008)

Furthermore, the concept of social sustainability by Goodland, 2002; Turkington & Sangster, 2006; Chan & Lee, 2008 and social SD Vavik & Keitsch, 2010 have been discussed.

3.1.2 Economical

The strategies of economic sustainability are used to encourage the best utilization of socio-economic resources. A sustainable economic structure suggests a fair distribution and proper allocation of resources. Moreover, it also encourages a proper and responsible use, so that it brings long time profitability. If the business is profitable and stable then it indicates a continuous process towards profitable business. On the other hand, we can get a clear idea regarding sustainability, when we focus on social and environmental issues and profitability will also follow. Besides, social powers play a vital role in consumer behavior and employee performance, while environmental power like energy efficiency and pollution that can reduce waste directly. Basically, economic sustainability means that the company will not make any decisions for the future generations while it is making their business profit. (Definition and pillars of sustainable development)

3.1.3 Environmental

Environmental sustainability means when all kind of activities, systems, and processing decrease environmental pollution from an organization's facilities, production, and operations. We are living in our natural resources but all our natural resources are not eternal, that's why we should make sure that the resources we using that rate should be in sustainable rate. It's a matter of thinking that environmental sustainability should not be messed with total sustainability, which also needs to balance social and economic factors as well.

According to Herman Daly:

- The level of harvest should not exceed the level of regeneration while considering renewable resources.
- For population, the level of waste production should not exceed the assimilative ability of the environment.
- For nonrenewable resources, the reduction of the renewable resources should require equivalent development of renewable substitutes for those resources.

Now a day many companies are concern about sustainable development and they are already taken or taking their right steps.

According to *Wal-Mart*, they have taken environmental initiatives by increasing imports from green and yellow factories, a goal of 0% of waste, plastic bag reduction, and proper steps to reduce carbon footprints by managing energy consumptions. *Nestle* also is taken some initiatives for the sustainable environment. They figured out four main areas to manage their environmental sustainability: water, agriculture, raw materials, manufacturing and distribution, and packaging specific to their food and beverage business.

Vanguard is an accessories company. They manufacture high-quality photo and video accessories. They have also a mission for the sustainable environment. Vanguard is ISO-14001 certified company, their products are RoHs compliant and the result of the proactive plan and maintain environmental management system, demonstrated to conformance to set policies and the meeting regarding environmental laws and regulations. *(Definition and pillars of sustainable development d)*

3.2 Adverse effects of Plastic bags

It is known that about 5% of world annual petroleum is used to make plastic directly. Moreover, another 5% of petrol burned to the fuel in the process. However, as plastic has much more negative effects on the earth surface, so surely it can be said that non-renewable plastic products are responsible for climate change every year. Although, the truth behind plastic is still much more complicated. Present days plastic is used in automotive instead of steel. Plastic makes cars lighter and faster. On the other hand, plastic bottles are used instead of glass, because of lightweight and non-fragile purpose. Although, plastic provides some benefits but according to the chemical industry, the use of plastic more than compensates the green house gas emissions of their manufacture. Even, if we agree that, plastic is much more carbon efficient than other materials in automotive or transportations, but we are still producing a vast amount of carbon emissions. Certainly, every year plastic use discharges minimum 100 million tons and maximum 500 million tons of carbon dioxide in our green atmosphere. According to Environmental Protection Agency (EPA), by producing every ounce of polyethylene it emits one ounce of carbon dioxide. PET plastic is widely used to make plastic bottles. The ratio between CO2 emission and plastic producing is closely 5:1whereas, every year we consume about 100 million tons of plastic. Natural gas and petroleum are widely used to make plastic. Both are highly considered for producing CO2. *(Plastic bags and climate change and destroying environment in variety of ways)*

Indeed, plastic bags and climate change are closely related each other in a variety of ways. From air pollution to ocean toxicity, plastic bags play a huge role to our ecosystem disruption. Every year only the USA manufactured approximately 30 million plastic bags where they use 12 million barrels of oil to manufacture. On the other hand, plastic bags are wasteful after used and unnecessary way to exhaust our oil supply and contribute spread CO2 through our atmosphere. In contrast, throwing plastic products in the ocean is another great reason of ecosystem disruption. Approximately, 100,000 ocean animals and birds die every year from suffocating on or consuming bags. While the number seems very little when we consider the extreme impact of spoiled bags that break up little pieces and wash into our waterways. Therefore, these little pieces of plastics are gathering at an alarming number in our oceans. Moreover, many people dispose their bags properly but it is still a threat to our ecology. For example, dioxin and other harmful toxic leach out of lands and contaminating waterways to ocean and lands as well. In addition, after using our bags we simply throw away, we think that by throwing away we are free from it but ultimately we are not. We should remember one thing that, there is no 'away'. Because, every bag we have ever thrown away is stored somewhere in our lands or waterways like canals, lakes, rivers, and oceans respectively. *(Plastic shopping bags and environmental impact (n.d)*

Plastic bags have a destructive impact to our food habit as well as aquatic animals food habit. Because it's toxic is polluting our crop fields as well as all our waterways. It doesn't matter how much we try to keep our cities clean many of the plastic shopping bags end its life as a waste. Despite, many of our cities streets are full of plastic shopping bags especially developing, under developing and poor countries. While developed countries are trying to keep their cities clean but still there are plastic bags. The reason behind is we are continuously producing and using plastic bags and it's in alarming rate around the world. Moreover, some researchers have done a research in 2004. They have found that, 6 pounds of plastic for each pound of plankton in the north pacific gyre. Besides, after 4 years in 2008 some researchers from Algalita Marine Research Institute found terrifying amount of plastic. It was 42 pounds of plastic for every pound of plankton. It increased 8 times by 4 years. Meanwhile, a research has been done by world economic forum and Ellen MacArthur Foundation. They claim that everyday about one dump truck full of plastic enters to the oceans. Moreover, scientists also say that there will be more plastic than fish by 2020, if we continue using plastic such a massive amount. *(Plastic bags and climate change and destroying environment in variety of ways)*

In contrast, plastic creates a massive breathing obstacle for aquatic animals and thousand of aquatic animals are dying each year. Besides, many more are gradually affected by dioxin and other harmful toxic. Meanwhile, dioxin is an endocrine disrupter and its called gender-bender pollutant. Basically, it is responsible for gender mutation in fish and land animals that eat fish such as sea bass, seals, even polar bear. On the other hand, we also eat fish so it's getting more dangerous for human life as well. The plastic in waterways is affect as all. Certainly, it will enter our food chain gradually and contribute to extinction.

Photo: Seas of Plastic. Source: Adverse effects of Plastic, a

Photo: The big plastic oceanic mess. Source: Adverse effects of Plastic, b

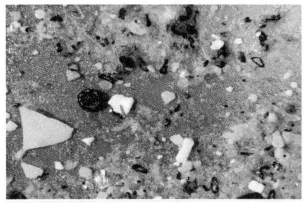

The Fallacy of Cleaning the Gyres of Plastic With a Floating "Ocean Cleanup Array" Photo: Stiv Wilson. Source: Adverse effects of Plastic, c

Photo: Kristine Lofgren, Albatross chick, Source: Adverse effects of Plastic, *d*

Photo: Chris Jordan Source: Adverse effects of Plastic, e

Figure 5. Adverse effects of using Plastic

3.3 Sustainability and Ecological aspects of packaging

Due to global warming, big and small companies are much more aware of saving environment. Therefore, companies are trying to design their packaging system on sustainable and environmental friendly ways. Below some aspects shortly described: *(Definition of Sustainable Packaging 2011)*

1. It will be beneficial, safe and healthy for individuals through its lifetime.
2. Its necessary to meet market criteria for performance and price.
3. Renewable energy should be used for its manufacturing, sourcing, transporting.
4. The use of renewable or recycled source materials should be improved.
5. It is necessary to use clean technology and best practices while manufacturing.
6. Through out its life cycle it uses all healthy materials.
7. It is designed to optimize in a physical way.

3.3.1 Brief explanation of points 1-7

Due to globalization, big companies started their strategies and business activities one part to another part of the globe and are increasingly being held accountable for actions resulting in negative social and environmental consequences. The raising emphasis on corporate citizenship, accountability and transparency report reflects an emergence of corporate social responsibility and sustainability. Renowned companies are implementing holistic sustainability by doing benchmark, measure, and tracking progress across a massive range of environmental and social impact categories.

The rapid growing packaging industry all over the world has been estimated about 430 billion dollars in 2009 and more than 5 million people are linked with this industry. For a product, packaging has so many benefits and it varies from the protection, preservation, transport of products and foodstuffs. On the other hand, packaging performs a vital role in product marketing, specification. Besides, it also informs consumer and educates the consumer about the products as well. It is believed that through intelligent eco packaging process, we can save our ecology and society. The life cycle of eco packaging paly a role for economic and environmental development. Moreover, it's a challenge for many countries to manage their waste ecological friendly way.

The ecological waste process supports persons and communities, through the creation of gainful employment, development of recovery infrastructure, conservation of resources and improvement of ecological performance. Besides, responsibility, accountability, and equitable wages all parts are part and parcel of sustainable development in the corporate social system.

The important part of sustainable development is economic growth and prosperity. The world population is growing every year and it is predicted that population will raise 33% by 2050 according to UN 2010. The aim of ecological packaging is to increase economic growth by providing packaged goods without creating the negative impacts that are traditionally associated with packaging or the other related process. Making profitability is a fundamental part of the sustainable business process. By maintaining the cost of packaging, procurement, production, product delivery with the required functionality and presence is an element of a profitable business. By doing an observation of Sustainable Packaging Coalition SPC membership, they have identified that the global packaging cost has become more and more complicated as costs that have traditionally been assumed by socially or environmentally. Eco packaging design always tried to minimize the cost by maintaining 100% eco package life cycle.

A huge part of the population is using natural energy resources every day, as a result it causes different types of problem in our environment such as – climate change, acid rain, mercury deposition, photochemical ozone, particulates. According to scientists, only renewable energy and eco-friendly products can save our environment from such calamities.

Big companies are making different strategies to convert renewable energy. Transportation is the big consumer of fossil fuel. By improving their fleet performance through an optimized distribution company can be benefited by fuel efficiency and it will minimize companies fixed cost as well. On the hand, companies are eagerly interested, to use alternative energies instead of fossil fuel, such as- hybrid, electric etc. These kinds of activities are encouraging the present market to the sustainable energy rapidly. By managing properly to our renewable materials or bio-based materials, we can create a continuous flow of sustainable energy that will ensure an availability of sustainable materials for our future generation.

Using bio-based products or recycled products can support the development of ecological packaging by developing its environmental data and supplying source of future packaging materials. A brilliant way for developing ecological packaging is optimizing the use of bio-based or recycled materials. Moreover, based on price, enough quantity and performance of bio-based or recycled materials affect the feasibility of integrating them into new packaging design. Materials and technological development that positively encourages these factors of sustainability that improves practically of their use. On the other hand, the packaging design and the usefulness of recovery system are closely connected with the sources of recycled materials. The need for recycled materials and the creation of end market is a key way to make energetic recovery and recycling industries needed to provide them. One of the most prime parts of recovered material is quality.

Green production is very important for the continuous application, which integrated ecological strategy to enhance overall efficiency and decrease risk to the human and the environment. This strategy involves maintaining raw materials, water, and energy, removing toxic and dangerous raw materials and also decreasing the quantity and toxicity of all emissions and waste while processing the productions. Green production signifies ecologically responsible practice and applies to any industrials activities including the packaging production. In packaging production manufacturing, it uses a major amount of water, energy, and materials, which create an impact on the environment. Green production search to implement ecological practices and modern technologies to reduce the ecological impact of manufacturing methods including and toxic used or emitted. Now a day, ecological friendly methods are persuading to decrease emissions, energy used, and wastes respectively. Inspiring manufacturer and suppliers to make sure that their all activities meet green production and highest practice standards, and assemble all prospects for responsible manufacturing. Moreover, worker safety is the major part that is linking to manufacturing performance to ecological packaging. Besides, the method will reduce the cost; improve quality and long-term profitability by reducing risks and improving fulfillment.

Sometimes, while manufacturing packets it contains certain chemicals. Through its life cycle, it releases these harmful chemicals, which create a negative impact on our eco-system.

' It is not possible to repeat too often that waste is not something which comes after the fact. Picking up and reclaiming scrap left over after production is public service, planning so that there will be no scrap is a higher public service.' -Henry Ford 1924

Generally, companies are trying to design packet to meet their critical cost, performance, marketing, and regulatory requirements.

4 CASE COMPANY PROFILE AND BIOBAG (SHOPPING BAG)

4.1 Introduction

BioBag® World or BioBag® International AS is a Norwegian Manufacturing company. The head quarter of this company based in Askim, Norway but it has also a head quarter in the USA. The company has a production site at Dagö an island off the coast of Estonia. The company has subsidiaries in Sweden, Finland, Denmark, Ireland, Australia and USA and partners representing BioBag international in many countries. The company is manufacturing not only shopping bags but also biodegradable mulch films, which are widely known and used for agriculture. On the other hand, the company offers a wide range of products to collect organic waste from the different place such as municipalities, commercial kitchens, and institutions. Every product made by BioBag® are ecologically sustainable alternatives to regular plastic products. A massive amount of organic waste is created by the HORECA industry, which is a part of BioBag®. The created waste is considered as huge resources for the company and that should not be sent to landfills. Besides, composting is the best way to handle the organic wastes. The cheapest and hygienic way to collect waste is to use with BioBag bags and liners. Today HORECA uses a broad range of food waste bags, sacks, and liners that are ecofriendly. *(BioBag, 2016)*

4.2 BioBag® (shopping bag)

Biobag is an eco-friendly shopping bag, which is manufactured by a company named BioBag® World. The special feature of this bag is that the materials are used for manufacturing BioBags are 100% compostable and biodegradable. The company uses all kind of eco-friendly raw materials for manufacturing their bags and other products. This bag is really convenient and easy to use. Today we are living an era which is full of plastic shopping bags. We have already known that plastic shopping bags are terribly dangerous for our ecology. Everyday we are using million of shopping bags all over the world. So the main purpose of this company is to manufacture 100% environmental friendly shopping bags so that in near future we can keep our environment clean and secure for our future generation. BioBag shopping bags are mainly found in stores and super markets such as S-group stores in Finland. BioBags are made resin derived that is produced from plants, vegetable oil, and compostable polymers. Moreover, according to the company, bags are made from compostable resin named Mater-Bi®. Basically, Mater-Bi® is produced by Novamont. Novamont is an Italian company dedicated to eco-friendly alternatives to polythene-based products. There are many countries around the world trying to minimize the use of plastic bags. BioBag shopping bags are 100% eco-friendly. By giving support to sustainable development for the future, we need to use materials containing renewable resources, which will minimize negative ecological impact. According to the life cycle of the BioBag shopping bag, it can be considered as a natural part of the environment. *(BioBag, 2016)*

4.3 Environment (we care) Compostable and biodegradable

The company BioBag is continuously searching for the best raw materials and manufacturing systems available across the world. BioBag is strongly trying to use renewable resources in their products as much as possible. They think that this process will reduce environmental pollution. Biodegradable and compostable are two words that are strongly connected with the environment. These two words are usually used while describing organic materials breaking down in a specific environmental friendly system. The definition for biodegradable is 'The materials which are able to undergo biological anaerobic-aerobic degradation heading to the production of CO_2, HO_2, methane, biomass, and mineral salts depending on environmental conditions in the process. Microorganism plays a vital role in biodegradation. Microorganisms mostly found in the environment and fed by organic waste. *(BioBag, 2016)*

All BioBag products are certified as compostable. On the other hand, unlike compostable, the system biodegradable refers very little as everything is biodegradable given time. Therefore, it is very crucial to identify the environment where the process can take place properly. Besides, composting is a process that helps to break down organic waste by microbial digestion to make compost. Meanwhile, compost has so many benefits; such as it helps the soil to improve and increase its fertilizer power. A proper composting process needs the exact amount of heat, water, and oxygen to make organic waste as compost. Every organic waste contains million of microbes that consume the waste and this process turns into compost. To achieve a certificate of compostable product that company produce must meet all the requirements in the European norm EN 13432 and US standard ASTM D6400. *(BioBag, 2016)*

4.4 Life cycle of Biobag

The concerns of environmental issues are growing rapidly and this growing interest is creating a new market for new products and services that are authentically eco-compatible. This new environmental compost has created a series of instruments that measure, control and check the ecological impact. Moreover, Mater-Bi and Novamont have always played an important role in the promotion for life cycle analysis and also EPD (Environmental product declaration). Besides, these factors are reflected as verifying the genuine environmental benefit of using biodegradable plastic polymer. *(BioBag, 2016)*

Life cycle analysis is a process for analyzing manufacturing activities, products and services, from a universal viewpoint. The simple way to study the overall production system is to go it step by step and examine its ecological performance. The whole path covered by the raw materials goes through from production transformation, logistic processes until the final action. This approach is termed from cradle to grave. The character of life cycle analysis is essential in recognizing the manufacturing process that has a massive ecological impact and indicating the routes for development in order to maximize the progressive effects and decrease the harmful effects on the environment to the lowest. There is another essential part called EPD (Environmental Products Declaration). The EPD is a significant voluntary tool for qualifying companies, which are ready to take an important role in managing ecological factors.

Today, the life cycle thinking has become crucial in evaluating the ecological impact of a product or service. Life cycle thinking contains the entire product value chain and identifies where changes and innovations can be made to it. *(BioBag, 2016)*

By studying the supply of raw materials, production units of a product in the same framework gives enormous potential for many new products. On the other hand, LAC (Life Cycle Assessment) is controlled by ISO 14040 and ISO 14044 standards. Below is a visual summary of the life cycle of BioBag based on the text above.

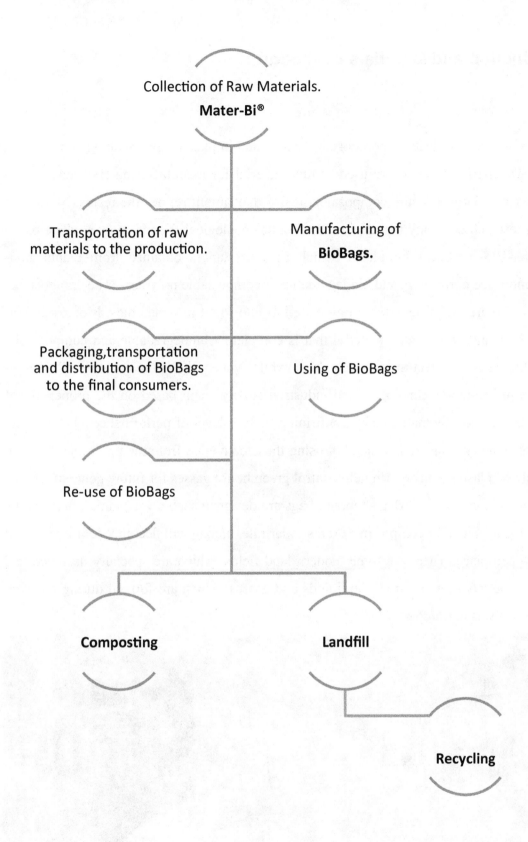

Figure 6. Life cycle process of BioBag® (picture made by author)

4.5 Production and Materials collection

Mater-Bi®

BioBag is known for products of compostable bags and bio plastic films making company on the market. Mater-Bi® products are used as raw material for manufacturing BioBags. Mater-Bi® is the world's largest family compostable resins. Novamont refines the resins. Novamont is an Italian research company that decided to produce ecological friendly polyethylene based plastic. Mater-Bi® produces bio-plastic which uses substances obtained from plants, non-genetically modified cornstarch, and biodegradable or compostable polymers. The compostable polymers are acquired from renewable raw materials and fossil raw materials. Moreover, Mater-Bi® is producing eco-friendly material that is completely biodegradable and compostable. More than 200 patents protect the products. Meanwhile, Mater-Bi® is the outcome of the non-stop struggle to use new technologies and industrial supply chain based on the proper use of renewable energy and raw materials by confirming the top level of performance. The process of using renewable raw materials instead of using those converted from the traditional fossil in the chemical and plastic industry can help control green house gasses for future generation.

According to the company, BioBag shopping bags are designed such a way so that they can be completely degradable after composting. At a similar rate, biobag will decompose if it is placed in a turn or open place. There are some modern land fields, which are specially designed and made to save the environment from the liquids and gasses, which are formed during the very slow breakdown waste. *(BioBag, 2016)*

Figure 7. Mater-Bi®- Raw materials of BioBag (Source: Mater-Bi® website)

4.6 Final products

Products of variety, BioBag is believed as the most comprehensive in the compostable product at market place. The main focus areas of the company are high –quality products, superior customer service, and their unique customization capabilities. The BioBag brand is already accepted as the best compostable product on the market.

Basically, BioBag shopping bag is made for long time use, its strong and elastic, it can be used about 10-12 times before composting. Shopping bags are available 8 different colors with best print quality. There are 4 kinds of bags: *(BioBag, 2016)*

- ➢ T-shirt bags
- ➢ Grip hole bags with or without Patch reinforcement
- ➢ Grip hole bags on block
- ➢ Loop handle bag.

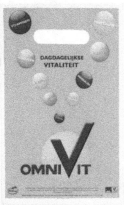

Figure 8. Samples of Eco-friendly shopping bags, photos: BioBag World (source: (BioBag, 2016)

5 CONSUMERS

5.1 Ecological consumers

The ecological consumers particularly described as one who adopts ecological friendly behavior or who buy eco-friendly products over the standards alternatives. They are much more self-controlled and aware of the environment. Moreover, they believe that anyone can be effective in environmental protection. They also believe that environment protection is not only a Government and business makers, environmentalists or scientists work but also all people from society can contribute to ecology. They are much open minded or tolerant towards new products or ideas. (ejeps)-5 (1), (Shamdasani at al, 1993:491)

According to marketing, there is always a certain amount of people or group of people who will use the products or services and these people or group of people are considered as consumers. It is vital to realize the behavior of consumers on any particular products or services. Each and every company try to understand consumer behavior like what makes consumers to select their products or services over a competitor's or why a consumer will buy their products of services. In order to get the answer or understand consumer behavior companies are always put much effort to analyze consumer behavior on their products and services. There are various definitions regarding consumer behavior-

"Consumers behavior echoes the totality of consumers' decisions with respect to the acquisition, consumption, and disposition of goods, services, activities, experiences, people and ideas (by) human decision making units"(Wayne & MacInnis 2009,p.3)

5.2 Difference between Bio Bag and plastic bag

BioBag shopping bag has greater positive impact on our ecology, while plastic bag has a lot of adverse effects. So here is some difference between BioBag and plastic bag.

BioBag shopping bags	Plastic Shopping bags
❖ It is 100% eco friendly. ❖ Can be used about 10-12 times. ❖ Medium lasting. ❖ Little bit costly than plastic bags ❖ It takes less time to become compost. Normal temperature about 90 days and hot temperature about 40 days. ❖ It is naturally compostable. ❖ It can be leak. ❖ Its weight is light to carry. ❖ Unusable bag does not make any harm to the environment. ❖ It can turn into fertilizer while it is composted. ❖ Using BioBag will save our ecology everyday and we can keep our environment green and safe for future generation. ❖ It has no negative impact to any species on this planet.	❖ It is not eco friendly. ❖ Can be used more than 20 times. ❖ Long lasting ❖ Cheap ❖ It takes many years to compost. ❖ It is almost impossible to compost naturally. ❖ It can be leak as well. ❖ Its also light weighted. ❖ Unusable bag has terrible bad impact on the environment. ❖ Its materials do not create any fertilizer. ❖ Using plastic bags in our daily life, we are just destroying our ecological balance. Similarly we are making a dangerously harmful environment for our future generation. ❖ Climate is changing because of ecological imbalance. ❖ Mass destruction is occurred because of natural disasters. ❖ It has vast negative impact on every species on planet earth.

Table 1:Difference between BioBag shopping bag and plastic shopping bag (source: made by author)

5.3 Creating awareness and encourage people to buy Environmental friendly shopping bags (BioBag)

The growing uses of natural resources are on a threatening level. This consumption affects to the natural environment and human health with pollution at dangerous levels. Therefore, eco-friendly products or applications are given top priority. The awareness of protecting the environment, known as green movement is highly appreciated by developed societies. Besides, they also adopted this eco-friendly movement and started to apply the procedure with the lowest potential to harm the natural environment. Manufacturing eco-friendly product is not sufficient by itself for a sustainable planet. The awareness for saving environment is also necessary. This matter is not only important for managing and marketing products but also consumers have the vital role in the operation of business. The environment lover consumers are already started to support by using the buying capacity and with their post consuming responsibilities. The movement of using eco-friendly shopping bags (i.e BioBag) that are not harmful to the environment, human and other species has become more and more popular to all ages of consumers. Certainly, it's the result of environmental awareness. (ejeps)-5 (1)

The BioBag has bright business opportunities in Finland. Finnish people are much more aware environmental issues. S-group customers are the biggest user of BioBag. Therefore, it can be said that S-Group consumers have knowledge about the relation between Eco-friendly and BioBag products. But it requires further information to deliver in a more effective ways. S-Group conceded one of the largest retail operators in Finland, S-group has a significant social responsibility for the ecology and considerable effect on consumer consumption pattern. " The impact of eco-friendly on the environment depends mostly on whether consumers buy eco-friendly products or not. The more consumers buy Biobag the more positive resulting impacts on our green environment. Therefore, the positive impact of BioBag on the environment can be evaluated indirectly on consumer awareness and acceptance of eco-friendly products on consumer behavior. (ejeps)-5 (1)

A consumer with an Ecological awareness can be defined as an ecologist who had grasped his or her self-efficiency against environment pollution and how has a consciousness of responsibility with the respect of future generation and the whole humanity in his or her use of resources. Besides, consumers who are alert with the ecological awareness can evaluate the existence of environmental reserves, their cost of consumption as well as the effect of this consumption to the environment and to themselves. (ejeps)-5 (1), (Babaogul and Ozgun, 2008)

There is a positive and significant relationship between environmental awareness and buying eco-friendly products. A survey conducted by Aslan on 400 university students at Kafkas University. From the survey, it was seen that students are conscious of eco-friendly purchase and use of products, which are harmful to the environment. Moreover, while purchasing the products they also notice about features of the purchased products, unnecessary packaging or wrapping, how waste terminate after using. This information has great importance even after purchasing the products to the students. (ejeps)-5 (1), (Aslan, 2007)

A survey made in Izmir founded that, when environmental pollution and environmental awareness rise, it strongly affect the consumer buying behavior. Besides, respondents were aware of the significance of recycling for protecting the environment and prevention of environmental pollution. (ejeps)-5 (1), (Aracioglu and Tatlidil, 2009: 435-661).

5.4 Environmental friendly products, price, promotion, place and purchasing

Due to globalization, environmental pollution increasing very fast and also leading to a huge reaction created against products threatening to the environment. When the threatening contains a product it makes consumers concern and influences the purchase decision of consumers. Besides, business has begun to manufacture eco-friendly products and create environment-friendly products policies. (ejeps)-5 (1), (Uydaci, 2002:113)

In 2009, Grail research conducted a survey on 520 American green consumers. The survey was made between 18-65 years old people, who are aware of environment-friendly products and who buy Eco-friendly products in their past life. According to the survey, consumers believe that eco-friendly products as those that decrease the impact on the environment such as energy-efficient, recyclable, natural or organic. Moreover, about 30% of the consumers think that minimizing water usage to be an eco-friendly practice, product labels and word of mouth are the major sources of information regarding eco-friendly products and companies for consumers. (ejeps)-5 (1), (Grail Research, 2010)

On the other hand, an affordable price will always encourage consumers to buy environment-friendly products. Certainly, the lower price is always a positive approach for a company when the demand of the products price is responsive. Besides, when the price is assumed as similar level, positive properties of the products about the ecosystem can be used as reasonable advantage component. If the price of the products is higher, then products promotion should be done decently and consumers should be ready to overpay for the products. In the situation, the main thing is the price. (ejeps)-5 (1), (Emgin and Turk, 2004)

Place indicates to the passage of products delivery. It can be from factory to store or store to final consumers or factory to final consumers. Choosing suitable products delivery passage is a strategic process. It determines the amount and the cost of products and market demand simultaneously. Generally, markets are determined to keep enough flow of their products. (ejeps)-5 (1), (Ricky & Ronald 2004, 319-399)

A decent presentation delivers the opportunities for the consumer to understand the relationship between business and environmental responsibility. The aim of the presentation is to create an image of ''environment-friendly business firm'' to the consumers and provide environmental messages to the consumers regarding the products. Therefore, to reach the goal marketing tools such as advertising campaigns, promotion, public relation and other tools should be adopted. This process requires both internal and external communication. (ejeps)-5 (1), (uydaci, 2002: 128)

5.5 Business and Eco-friendly marketing

In 1975 American Marketing Association (AMA) organized a conference on 'Ecological Marketing' and it was the very first beginning of 'Green Marketing' discussion in the seminar. The discussion was made about the impact of marketing on natural environment with the contribution of academicians, bureaucrats and their participates, and ecological marketing concept was also defined such as – adverse or positive impacts on environment pollution, energy consumption and consumption of other resources because of marketing. There are immense changes rising in the corporate business world regarding the responsibility towards the ecology and the society. Making profit is not only the main goal in the business strategies but also to keep consideration of eco-friendly and sustainability in the business agenda. The ethical code for corporate business is being green. Therefore, It is not longer existed that only making money for the company and thinking only the company. Therefore, companies are changing their old fashioned ideology and also inspiring other companies to be more and more Environmental friendly as well. Besides, Global warming, environmental issues, and social problems will be challenges for the future generation frontrunners for taking efficient and comprehensive decisions. In the way of making these decisions, the companies should give priority to the Environment same as profitability. (ejeps)-5 (1), (Cevreormen, 2010)

There are two purposes that green marketing serves such as:

> ➢ In order to develop goods that can appeal to the consumer, reasonably affordable prices and environmental friendly products causing minimal damage are required.
>
> ➢ In order to reflect and high quality image of high quality, environmental sensitivity and hence production of products compatible with environment are required. (ejeps)-5 (1) ,(Uydaci, 2002:85)

"The leaders of future generations are responsible not only for obtaining desired results but also for the impacts of the decisions on elements other than their own companies and markets"
(ejeps)-5 (1), (Bill Gates, Business news 2010)

5.6 Advertising

Advertising works as a bridge between consumers and products or services. It plays a phenomenal role to spread information regarding products or services and company to the consumers. It encourages people to buy the products or services. The ways of advertising are television and radio commercials, billboards, flyers; magazines, Internet and social medias and other methods of advertising are targeted to promote consumption in all its methods. There was a time when the main role of advertising was to make people buy more and more but present days that prototype tendency has changed. It is now responding to new demands from consumers searching for greater significance, transparency, and ethics. Therefore, commercial advertising can play an important and powerful force to promote eco-friendly consumptions in highlighting eco-friendly goods or services and convincing people to buy the products or services regardless of thinking about price. In OECD countries, they have some regulations for advertising regarding basic consumers protection.

Usually, it contains fair trading legislations and policies controlled by consumer protection agencies. Besides, it generally covers advertising claims made about the environmental, social or ethical attributes. For instance, almost all over the world automobile companies are trying to spread their concern regarding climate change and environmental features of their models such as Daimler Chrysler (Fresh Air), Honda (Safe and Environmental), Toyota (clean Air), Volkswagen (Save Fuel), Peugeot (Flower), Kia (Think Before You Drive) etc. Likewise, Different petrol companies are also creating a green image by advertising commitments to preserving nature such as BP (Beyond Petroleum) Shell (Gardener), and Total (Dolphin). Meanwhile, the company BioBag® is also promoting their products through all advertising media. Everyday consumers responses increasing and it indicates a bright and positive future for the company and environment. (ejeps)-5 (1)

6 RESEARCH METHODOLOGY

6.1 Research Method

The quantitative research method was used because this method displays the approximate picture of the research area. Answering questions regarding what, why, how many, global warming, using purposes of the shopping bag, Green environment etc. gives the chance the to identify the process and present descriptions which are the core part of quantitative research. Apart from that, there is no second chance to make an extra observation if a respondent misunderstands the questionnaire. The goal of this data analysis was to figure out the ultimate picture of the consumers' ideas, thinking styles, the eagerness of using BioBag, awareness of global warming and preference while using shopping bags in Helsinki region. The purpose of this questionnaire was to define how a consumer acts and think while buying shopping bag at the retail store.

6.2 Validity and reliability

Reliability is accomplished by precisely controlling all the data and related procedures. Validity shows the correspondence of the result with the reality. Validity can be divided into two parts like, external and internal. Internal validity is the research measure at the core point of the survey. On the other hand, external validity is associated with the generalization, in terms of how the findings of the study can be generalized. (*Gummesson 2000 and Yin 1994*)

Reliability suggests extending to which data gathering techniques or analysis procedure will yield consistency findings. Therefore, reliable observations yield the same results on other occasions and other by observers as well. It should also be cleared how the raw data was interrupted. (Saunders, Lewis & Thornhill, 2007:149)

Reliability also suggests to the quality of a measurement method, which delivers repeatability and accuracy. Impartial and objective means that every measurement has been taken in an impartial manner and without self-interest. (Kumar, 2005,6)

Validity is the ability of a tool to measure, what is design to measure. Besides, validity refers extending to which and empirical measure adequately reflects the real meaning of the hypothesis under consideration. (Kumar, 2005:153).

According to the Mason's (1996,147), ' you should ask yourself that, how well the logic has matched of the research method and research questions that you are asking and the kind of social explanation you are intending to develop. There was also a process, which is used in this research process. Various journals and relevant books, articles recognized in the international business field were used to gather information for the study.

The empirical part of this study was carried out via surveying 22 people with the help of a questionnaire (appendices) at Arcada campus and Hakaniemi tori on the month of October 2016. Questionnaires were given directly to the people and filled it out. The interview was conducted with the supervision of author. Only those who were interested to participated were interviewed, no pressure was applied any of the participators. Questionnaires were explained to the responding people in all the necessary contexts so that they could answer it properly. The main challenging part was to find out people who have time to fill out the questionnaires. As stated before that, there were no specific criteria for choosing people for interview. Every interested person was requested to take part. The respondents of this study were mainly, who are using shopping bag in everyday life, regardless of any varieties all types of people who were interested to participate in this survey were interviewed.

6.3 Results

After collecting all answers, it was sorted out carefully such as, unfinished and incorrectly filled questionnaires were rejected. Out of 22 filled questionnaires, 20 pieces were correctly filled and kept for analysis. Finally, results were analyzed through MS Excel.

6.3.1 Preference based on gender

The pie chart shows that there was 71% of people were the female who answered the questionnaire and 29% were males.

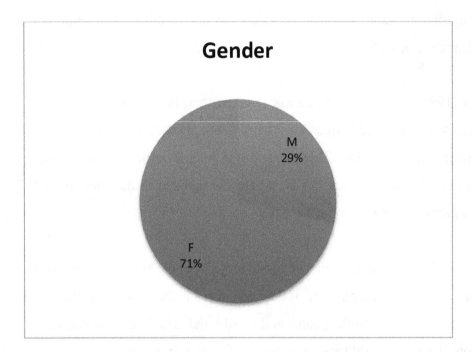

Figure 9. Shopping bag user gender (n=20)

6.3.2 Preference based on age group

Figure 12 explains that more than half of users age was 21-31 years old. About 4 persons were 31-41 and at least 1 person was 41 and above 51 years old.

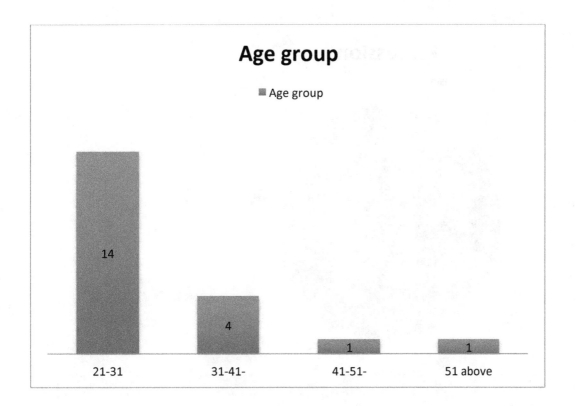

Figure 10. Age group of uses shopping bag (n=20)

6.3.3 Profession

According to figure 13, it can be said that 62% of the users were students, 29% were employees and 9% of the users were businessmen.

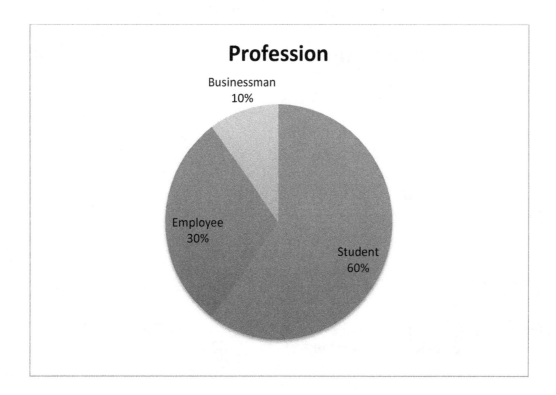

Figure 11. Profession of the users (n=20)

6.3.4 Plastic bags users/month

In figure 14, it illustrates, the use of plastic bags is high. More than 10 bags were used per month. 12 persons said about that. About 4 persons use 6-10 bags a month and rest of the users uses 1-5 bags per month.

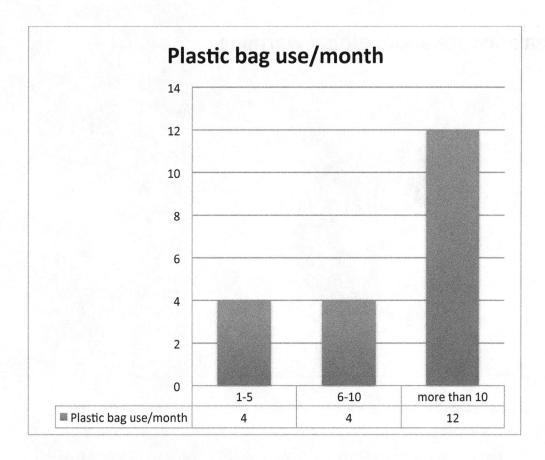

Figure 12. Monthly user of plastic bags (n=20)

6.3.5 Idea and care about global warming

Figure 15 shows that about 95% of people have knowledge about global warming. On the other hand, 5% of people don't care about global warming and they have no idea about that.

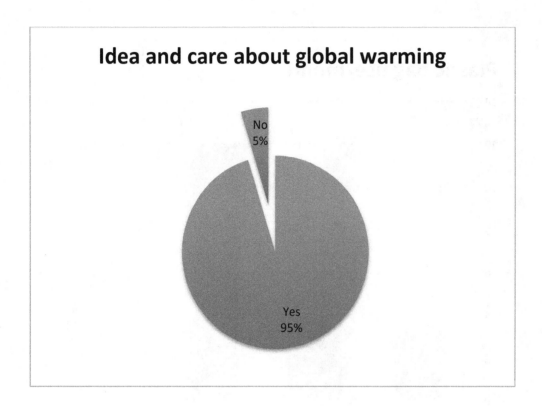

Figure 13. Idea and care about global warming (n=20)

6.3.6 Familiarity of Eco friendly bag

This chart says that approximately 90% people have knowledge about eco bag and its positive sides.

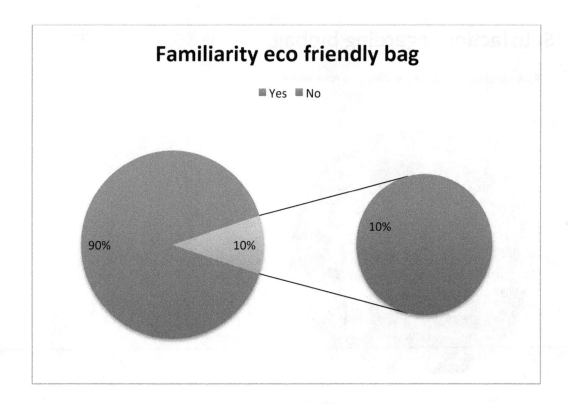

Figure 14. Familiarity of Eco friendly bag (n=20)

6.3.7 Satisfactions about using BioBag

Figure 17 displays that; about 50% of the users were quite satisfied whereas, 45% were fully satisfied and 5% didn't give any comment. So finally it can be said that BioBag has positive future.

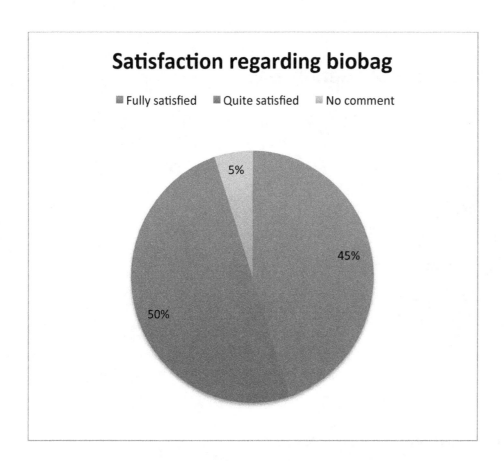

Figure 15. Satisfactions about using BioBag (n=20)

6.3.8 Plastic bag after use

Most of the plastic bags reused for different purposes. Only 15% bags are used only one time. It doesn't matter how many times people are using but in the end, it is thrown away.

Figure 16. Use of Plastic bags (n=20)

6.3.9 Business Idea

It was asked to the people that do you think the substitution of the plastic bag by BioBag is a good business idea? Nearly 16 persons were completely agreed and 3 of them were partially agreed just 1 person has no idea about that. According to the graph, it is clearly seen that BioBag has a good future in the market.

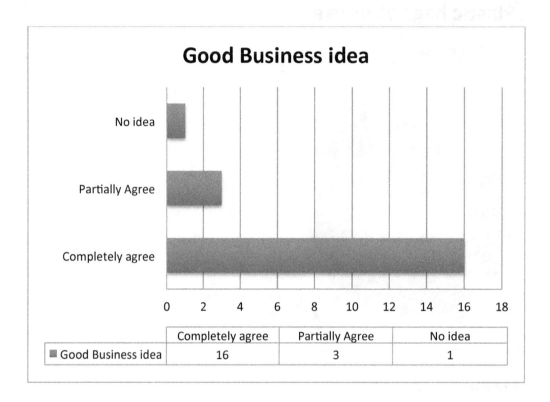

Figure 17. Business Idea of manufacturing BioBag (n=20)

6.3.10 Eagerness of using BioBag in everyday life

Figure 20 illustrates that 90% of people were agreed to use biobag in their daily life while only 10% people were not interested either they use the plastic bag or eco bag.

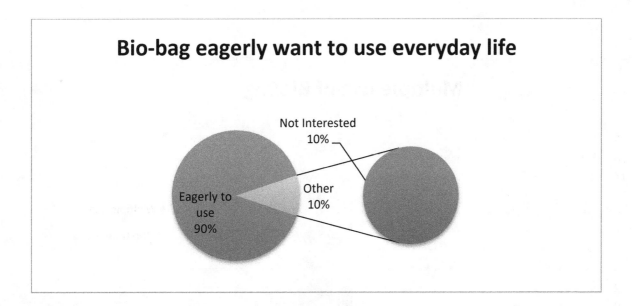

Figure 18. Eagerness of using BioBag in everyday life (n=20)

6.3.11 Multipurpose use of Biobag

According to Figure 21, it can be said that people are more interested using BioBag in various ways. It's a really good side of Biobag for future business. About 18 persons were interested using it more than one time. Only 2 persons wanted to use it once.

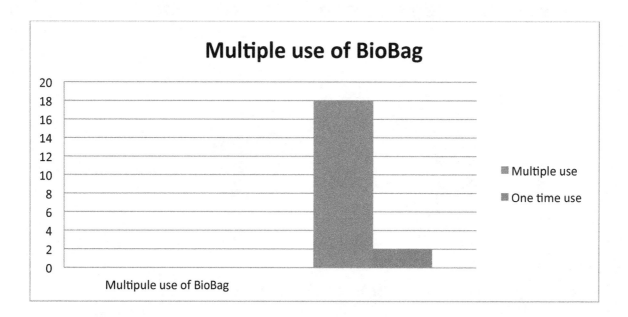

Figure 19. Multiple use of BioBag (n=20)

6.3.12 Summary of the results

The author tried to include respondents from different profession in this research. It could be better understood the opinion of different professions people who are in the different ages. Most of the time people use plastic bag by their needs. So sizes vary on their need as well. According to the research, it is seen that eco-friendly information can change consumer-buying decisions in a positive way. The survey also shows that the majority of the people are eager to use BioBag instead of plastic bag. BioBag has been successfully differentiated from other similar retail products according to the research. The majority of respondents agreed that there is enough difference between BioBag and the plastic bag. Only fewer respondents did not notice any different between two. While doing research work the author noticed that, most of the people were concern about ecology and its future. They are much more interested to use BioBag instead of plastic bag. So it is a positive attitude towards the BioBag industry from the consumer side. It is possible to change consumer behavior if they get right materials and right information. Respondents are willing to use the products, which are made by the ecological friendly material.

7 CONCLUSION

Due to the evolution of the technology, the change of human needs has increased enormously. The changing and increasing of human needs are vastly responsible for environmental pollution and it affects entire planet in a negative way. We have been in the search for ensuring our future and sustaining our life in appropriate condition. Consumers are not remaining insensitive to ecological problems such as environmental pollution and global warming. Consumers are being aware the product whether it is eco-friendly or not apart from price and quality features of the product. People are giving much more attention regarding environmental problems. Most of the companies have started environmental friendly activities and they are trying to reach 'Green Marketing' concepts to the consumers.

The core focus area of this thesis is the sustainable environment, creating awareness of global warming, the life cycle of BioBag (shopping bag), negative effects of plastic bags, positive sides of BioBags, eco-friendly products advertising, business opportunities of Biobag and consumer behavior in the theoretical framework. The thesis contains a case company description. The study also contains an interview with case company CEO. Moreover, a survey has done by the help of some questionnaires.

7.1 Summary of the theoretical framework

The meaning of sustainable development is to use resources without damaging the nature for the future generations. It requires habit in all stage to global. The factors like economic, social and environmental combine together in sustainable development. It is worthwhile to make policy in this sector. Global warming is an immense threat to our planet earth. By ensuring a sustainable environment, it is possible to minimize global warming. In 2009, international climate change agreement has made by EU and in the agreement, it was committed reducing overall emission by 2020. Which will be 20 percent below levels of the year 1999. There are some parts that belong to sustainable development such as waste hierarchy, recycling; the prevention of waste and reusing of materials and the integration of life cycle viewpoint is also part of sustainable development.

The viewpoint of lifecycle starts from raw materials, production to final disposal. The polymerization and ethylene are used to produce plastic bags. Besides, a gaseous hydrocarbon existed in petroleum or oil. Plastic is a silent killer of nature. It takes about 400- 1000 years to vanish, after throwing it to nature. There are some other ways to vanish plastic but it's also harmful to the environment.

Mater-Bi® is a natural component which is used for making all BioBag (shopping Bags) and it is 100% biodegradable and compostable. It is well-known material that helps sustainable development. It assembles ecological demands with those of agriculture and industry. The Mater-Bi® is developed by Novamont. It is the genuine response to the demand for convenience products, which have some environmental impact. It originates from renewable resources of agriculture basis. It decreases gas emissions and the utilization of energy and non-renewable resources. It has a virtuous life cycle such as agriculture raw materials return to earth through the process of biodegradable and compostable. Moreover, it's also certified under very strong regulations on environmental matters. Novamont and Mater-Bi® can be a good example model. Novamont produces and sells different lines of biopolymers for a different manufacturing method and all with the Mater-Bi® trademark. According to the research, the majority of consumers obtain their BioBag facts and information from company publications and the Internet. This is a positive indication that consumers seek out information willingly when in need of BioBag® information. However, there is a sufficient amount of success also in the advertisement, being one of the most popular sources of BioBag information.

7.2 Suggestion

The research results also show positive indications regarding the eagerness of consumers to adjust to more sustainable consumption pattern and their eagerness to pay extra for environment-friendly products. The BioBag® company should always keep finding new and creative ways to reach its costumers. The research result also indicates that a lack of marketing communication regarding BioBag products. The outcome clearly shows that S-Group's customers already have basic knowledge about BioBag and environmental issues, but they need further information, delivered in a more effective and easy way to understand. The image of eco-friendly and sustainable consumption must be transformed in the mind of customers, making it mainstream thought and part of everyday life. The confirmed and effective marketing methods used promoting other areas of retail business should be applied to BioBag products as well. Marketing section could such as promote sustainable consumption as a fashionable and fun product group for all especially children while highlighting healthiness for women as well. The primary financial obstacle should be considered, but as BioBag products consumption increases, the price will decrease. On a larger scale, sustainable production is an important part of the future. Besides, the company can take advantages and it can arise to experience positive results. Meanwhile, every company needs innovation to endure in the business. Every time competitors are introducing new ideas and products. Innovation is not a single process, a company needs to run through a series of activities where the company requires to follow different steps to reach its goal and make company profitable and Eco-friendly simultaneously. However, a new product does not mean the product must appear completely new. It performs a verity of purposes depending on what is seen to be a strategic imperative. Precisely, consumer behavior considered a vital part to any kind of business. Most of the people like to keep their products separate in order to category of products and they want to use easy and comfortable bags. Now a day, some countries like Australia; France, Norway, Denmark etc. are trying to ban the use of plastic bags. Big companies like IKEA are also going to stop using plastic bags soon. According to the research, it is clearly forecasted that there is a vast opportunity in future for BioBag® company.

7.3 Suggestion for future research

Finally, it can be also said that this same area of research could be carried out in different perspectives such as organizational perspectives, analyzing how companies are currently emerging sustainability issues into their strategic plans and organizational culture. The study could be successfully accomplished using the qualitative method on various levels of an organization or several different organizations.

REFERENCES

Adverse effects of Plastic a:(n.d) retrieved from http://www.gard.no/web/updates/content/20856696/seas-of-plastic-the-ocean-cleanup-%20http://truthfrequencyradio.com/the-ocean-cleanup-plastic-recycling-alternative-energy-environment-and-imagination/ Retrieved 15 July 2015

Adverse effects of Plastic b, (n.d) retrieved from http://truthfrequencyradio.com/the-ocean-cleanup-plastic-recycling-alternative-energy-environment-and-imagination/ Retrieved 15 July 2015

Adverse effects of Plastic c (n.d) retrieved from http://inhabitat.com/the-fallacy-of-cleaning-the-gyres-of-plastic-with-a-floating-ocean-cleanup-array/ Retrieved 15 July 2015

Adverse effects of Plastic d (n.d) retrieved from http://alloneocean.org/clean-up-stations/ Retrieved 15 July 2015

Adverse effects of Plastic e (n.d) retrieved from http://inhabitat.com/donate-to-suck-millions-of-tons-of-plastic-out-of-the-worlds-oceans/ Retrieved 15 July 2015

BioBag (2016) Retrieved from https://biobagworld.com/ Retrieved 25th September 2016

Boztepe, A, (2012), *European Journal of Economic and Political Studies* (ejeps)-5 (1) retrieved form http://oaji.net/articles/2016/3041-1455609133.pdf Retrieved 25th September 2016

Cause and effects for global warming, (n.d) retrieved from http://timeforchange.org Retrieved 20th September 2016

Definition and pillars of sustainable development b, (n.d) retrieved from http://www.circularecology.com/sustainability-and-sustainable-development.html#.WSyc1SN94b3

Definition and pillars of sustainable development c retrieved from https://www.frontstream.com/the-three-pillars-of-sustainability/

Definition and pillars of sustainable development d, (n.d) retrieved from http://www.vanguardworld.hk/photo_video_hk/company/sustainability/ Retrieved 2 October 2016

Definition of Sustainable Packaging (2011) retrieved from http://sustainablepackaging.org/uploads/Documents/Definition%20of%20Sustainable%20Packaging.pdf Retrieved 20th September 2016

Erdos, J. (2012) Why Choose BioBag as Grocery Bags? Go Green with Durable Shopping Sacks to Reduce Waste Retrieved from http://www.huffingtonpost.com/2012/04/06/reusable-grocery-bags_n_1409065.html 30th August 2016

Fossil carbon emission, (n.d) Retrieved from http://www.solarnavigator.net/fossil_fuel.htm Retrieved 20th September 2016

International Institution of Sustainable Development 2010, Definition and pillars of sustainable development a, (n.d) retrieved from http://www.iisd.org/topic/sustainable-development Retrieved 20th September 2016

Mater-bi (n.d) retrieved from http://www.novamont.com/eng/mater-bi Retrieved 20th September 2016

Plastic shopping bags and environmental impact (n.d) Retrieved from https://www.reusethisbag.com/articles/plastic-shopping-bags-environmental-impact.php Retrieved 30th January 2017

Prem, S., Ong-Chon-Lin, G., & Richmond, D. (1993) "Exploring Green Consumers In An Oriental Culture: Role Of Personal And Marketing Mix Factors, *Advances in Consumer Research*" 20: 491 Retrieved 25th September 2016

Shrum, L. J, McCarty, J. A & Lowrey. J. A., Tina M .(1995) "Buyer Characteristics of the Green Consumer and their implications for Advertising strategy" *Journal of Advertising*, 24, 2: 71 Retrieved 25th September 2016

Staley, S, (2005) Plastic bags and climate change and destroying environment in variety of ways Retrieved from https://alumni.stanford.edu/get/page/magazine/article/?article_id=30619 Retrieved 25th February 2017

Strategy of Sustainable Development- Basics of SD strategies (n.d) European Sustainable Development Network ESDN http://www.sd-network.eu/?k=, Retrieved 30th January 2017

Straughan, R, Roberts, J. A.(1999). "Environmental Segmentation Alternatives: A Look At Green Consumer Behavior In The New Millennium" *Journal of Consumer Marketing*, Vol:16,6 : 559-575 Retrieved 20th October 2016

Tilikidou, I, Delistavrou , A.(2008). "Types And Influential Factors Of Consumers Non-Purchasing Ecological Behaviors, *Business Strategy and the Environment*"61-76. Retrieved 26th February 2017

Trend of global gas emission (n.d), Retrieved from http://research.noaa.gov/News/NewsArchive/LatestNews/TabId/684/ArtMID/1768/ArticleID/10216/Greenhouse-gases-continue-climbing-2012-a-record-year.aspx Retrieved 20th September 2016

United Nations Framework Convention on Climate Change (UNFCCC) (2009) Sustainable Development and Global Warming retrieved from http://unfccc.int/files/meetings/cop_15/application/pdf/cop15_cph_auv.pdf Retrieved 20th October 2016

United Nations Framework Convention on Climate Change 2010. (2010) Press release about UN climate change Conference in Cancun delivers balanced package of decisions restores faith in multilateral process, Cancun, Mexico. 11 December 2010 Retrieved from http://unfccc.int/2860.php Retrieved 20th October 2016

APPENDICES

Questionaries Regarding BioBag Survey

Dear Reader,

I am a student of Arcada University Of Applied Sciences. I am writing my thesis about Business benefits of Ecological Packaging. I am writing specifically about BioBag (Shopping Bag). The case company of this thesis is BioBag World. I need your kind cooperation on this survey by answering the questionnaires'.

Finally, thanks your kind cooperation.

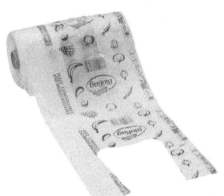

Please fill the form by × mark

1. Gender:

☐ Male

☐ Female

2. Age Group:

☐ 21----30

☐ 31----40

☐ 41----50

☐ 51 and above

3. I am a:

☐ Service holder

☐ Entrepreneur/Businessman

☐ Pensioner

☐ Student

☐ Unemployed

☐ other, please specify_____

4. How many Plastic bag you need in a month?

☐ 1- 5

☐ 6-10

☐ More than 10

☐ I am not using plastic bag

5. What do you really consider while using Shopping bag?
(Please tick one box for each characteristic) can mark multiple characters-

☐ Environmental Friendly material

☐ Cost Quality

☐ Durability and Easy to handle

☐ multipurpose use

☐ Usability (weight carrying)

☐ other please Specify

……………………………………

4. Do you have any idea about Global warming?

☐ Yes

☐ No

5. Do you care Green Environment?

☐ yes

☐ no

6. Are you familiar with eco-friendly shopping bag?

☐ yes

☐ no

7. How do you feel after using (BioBag)/ Environmental friendly shopping bag?

☐ Fully Satisfied

☐ quite satisfied

☐ Unsatisfied

☐ No comment

8. What do you do with plastic bag after using it?

☐ Reuse

☐ Bin liner

☐ throw to the Garbage Bin.

9. How many purposes do you want to use BioBag? (You can choose more than one)

☐ I can use it as a shopping bag for groceries

☐ I can use it for other shopping

☐ I can use it for travelling

☐ Any other, please specify

………………..

10. Substitution of plastic bag by BioBag - do you think it's can be a good business idea?

☐ Completely agree

☐ Partially agree

☐ Not agree at all

☐ I don't have any idea

11. Do you want to use Eco-friendly shopping bag (BioBag) Everyday?

☐ Eagerly to use

☐ Not interested

☐ None of above

12. If you have any more comment, please feel free to specify below-

_____

_____

CPSIA information can be obtained
at www.ICGtesting.com
Printed in the USA
BVHW010935180520
579872BV00012B/159